CURTISS P-40

by the Aeronautical Staff of AERO PUBLISHERS INC.
(In cooperation with The Air Museum)

Scale Drawings by Uwe Feist

Aero Publishers, Inc.
329 Aviation Road Fallbrook, California

Library of Congress Catalog Card Number

65 - 24307

Printed and Published in the United States of America by Aero Publishers, Inc.

CURTISS P-40 EVOLUTION

The original Curtiss XP-40, ordered July 1937, was converted from the tenth P-36A by replacing the radial engine with a new Allison V-1710-19 engine. It flew for the first time in October 1938.

This new liquid-cooled engine fighter had a casling radiator mounted under the rear fuselage, but the prototype XP-40 was later modified and the radiator was moved forward under the engine.

The P-40 design was based on a proven airframe already in production. Its low cost per production item and early delivery offered many advantages over other contemporary types.

Large orders were placed by the French and British governments for the P-40B model. The P-40 made history above the deserts of the Middle East while in use by the R.A.F.

One hundred P-40 B & C's were sent to China to serve with the A.V.G. Flying Tigers against Japan. The genius of Gen. Claire Chennault was born out by his tactful use of the P-40. He attained more success with his handful of P-40's than did the entire Army Air Force fighting at the same time in the Philippines. At the time of Pearl Harbor, the P-40B was our standard fighter and these Hawaiian-based fighters were sitting ducks at Hickam Field for Japanese attack planes. A few got into the air and one piloted by Lt. George Welch accounted for four Japanese aircraft. Thus the P-40 had the distinction of downing the first Japanese aircraft of the Pacific War.

The P-40D model had increased armament and later improvements went into the P-40E model which was ordered in very large numbers and delivered to AAF, RAF, and Russia.

The early model Allison-equipped P-40's had no Turbo-supercharger and thus its high altitude performance was very limited.

The P-40F model was born to remedy this situation and the U.S.-manufactured Packard V-1650-1 "Merlin" engine offered new high altitude capabilities with its two-stage supercharger. This model was called "Warhawk," while the early P-40B and C models were labeled "Tomahawk," and the P-40E, the "Kittyhawk."

Continual improvements increased production. The next large production order was for the P-40K model and the largest number were sent to China. Four hundred P-40L's were manufactured with Packard Merlin Engines like the P-40F model. The P-40M was the last of the heavy P-40 series. P-40N, introduced in March 1944, offered a new lightweight structure, with improved cockpit visibility. The first 400 P-40N models built were for high altitude work and were the fastest models produced. The P-40N was built in larger quantities than any other model P-40 produced.

One of the most interesting models produced in the entire P-40 series was the XP-40Q which featured an Allison V-1710-121 engine, four-bladed propellers, bubble canopy for better visibility, and wing radiators. Different modifications were made on this aircraft; however, it never went into quantity production, as the P-47 and P-51D were already in production for the Air Force.

The last P-40 came off the production line in December 1944, which was a record at that time for continuous production of a standard Army Air Force Fighter.

A total of 13,738 P-40's of all different models were built and when the end of World War II came, one American Fighter Group was still equipped with P-40 aircraft.

This Gallant War Horse fought on more war fronts than any other fighter: In China, Burma, Java, Alaska, Australia, New Guinea, Hawaii, Philippine Islands, Solomon Islands, in Russia, Italy, Turkey, and North Africa.

CURTISS P-40's

CURTISS P-40

Curtiss P-40B at Langley Field was with 8th Pursuit Group. Note special Tri-Color design on wheel.

Curtiss P-40C in Air Corps Service.

P-40D of Army Air Corps in Flight.

Curtiss P-40E with Tiger Shark Markings.

Curtiss TP-40N — Used as P-40 Trainer by AAF, World War II.

CURTISS MODEL 81-A "TOMAHAWK" II

In standard R.A.F. Western Desert Camouflage comprised of sand brown overall with irregular pattern of brown. Under surfaces were painted pale blue and the individual identification numerals were painted white; the Airscrew spinner was also white.

Engine:	Allison V-1710-33, 1040 H.P.
Speed:	352 M.P.H. @ 15,000 feet
Armament:	2-50 cal. M.G. in nose
	4-30 cal. M.G. in wings
Weight:	Gross, 7,650 lbs.
	Empty, 5,590 lbs.

Unusual front view of P-40N-35-Cu on U.S. Civil colors but retaining wartime Tiger Shark Marking.

Curtiss Hawk 87 of Royal Canadian Air Force.

An "Aleutian Tiger" — Curtiss P-40E. Flown in the Aleutian Islands by Maj. John Chennault, son of Gen. Claire Chennault, where he commanded a Fighter Squadron during World War II.

Maj. John Chennault's P-40E taxis out for Aleutian Fighter Patrol.

Close up of Curtiss P-40E showing engine and cowling details.

One of the first production P-40B's for U. S. Army during an acceptance flight.

Cowl close-up of P-40 shark insignia and engine exhaust stacks.

Detailed close-up of Curtiss P-40E landing gear which rotates 180° into wheel well when retracted.

P-40E landing gear. Tire and wheel were large 30-inch type for rugged dependability.

CURTISS P-40F "WARHAWK"

This Merlin-powered P-40 was flown off an aircraft carrier during the North African Invasion to support Allied Forces. The American Flag was painted on the sides of the fuselage for recognition purposes.

Engine:	Packard-Merlin V-1650-1
Speed:	365 M.P.H. @ 20,000 feet
Armament:	6-50 cal. Browning M.G.
Weight:	Gross, 9,350 lbs.
	Empty, 6,590 lbs.

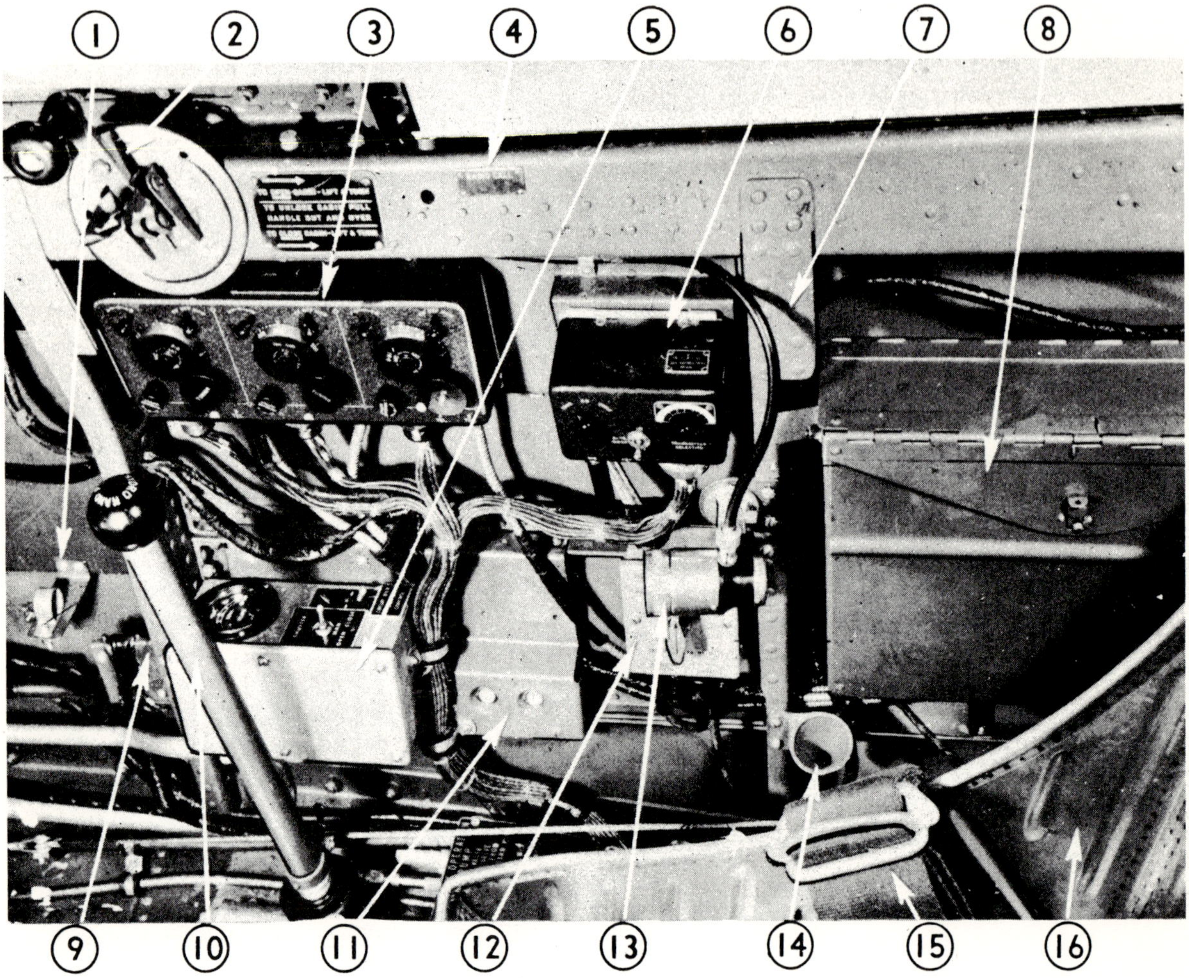

1. Bracket for BC-646-A control box
2. Cabin control
3. BC-450-A receiver control
4. Radio call plate
5. Cowl shutter control box
6. BC-451-A transmitter control
7. Station No. 4
8. Map case
9. Station No. 3
10. Auxiliary hydraulic pump handle
11. Radio crash switch
12. Filter switch box
13. Fluorescent spotlight
14. Microphone bracket
15. Safety belt
16. Pilot's seat

Right side view of P-40 Cockpit.

NATIONAL INSIGNIAS

Great Britain

Dutch New Guinea (1944)

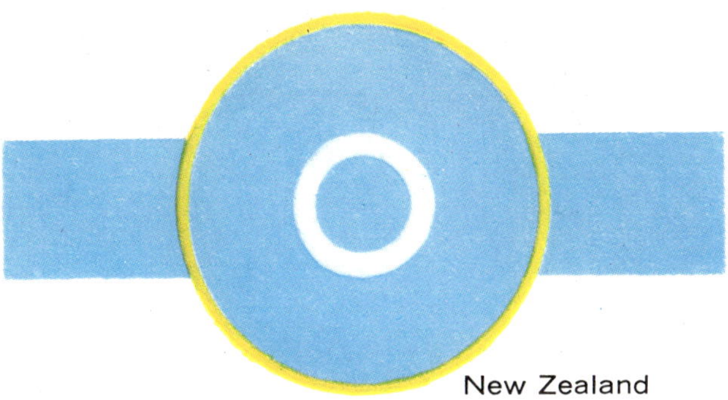

New Zealand

South Africa

Chinese Nationalist Air Force (1944)

CURTISS P-40E "KITTYHAWK"

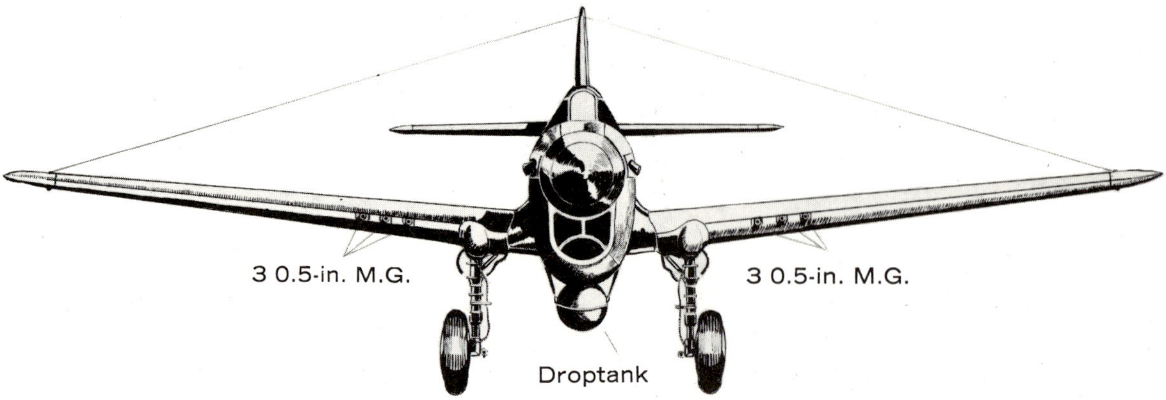

CURTISS P-40E "KITTYHAWK" of Major John Chennault

His Aleutian-Fighter Group carried the striking insignia of "Bengal Tiger" on the nose, instead of the Tiger Shark emblem made famous by his father, General Claire Chennault.

Engine:	Allison V-1710-39, 1150 H.P.
Speed:	350 M.P.H. @ 15,000 feet
Armament:	6-50 cal. Browning M.G.
Weight:	Gross, 8,840 lbs.
	Empty, 6,500 lbs.

美國空軍
來華助戰
仰我軍民
一體救護

國民政府
航空委員會
字第 W 187108 號
借用

Flying Tigers I.D. Flag attached to Back of A.V.G. Pilot's Flight Jacket. Later when Flying Tigers became part of 14th Air Force the American Flag was substituted for the Chinese Flag.

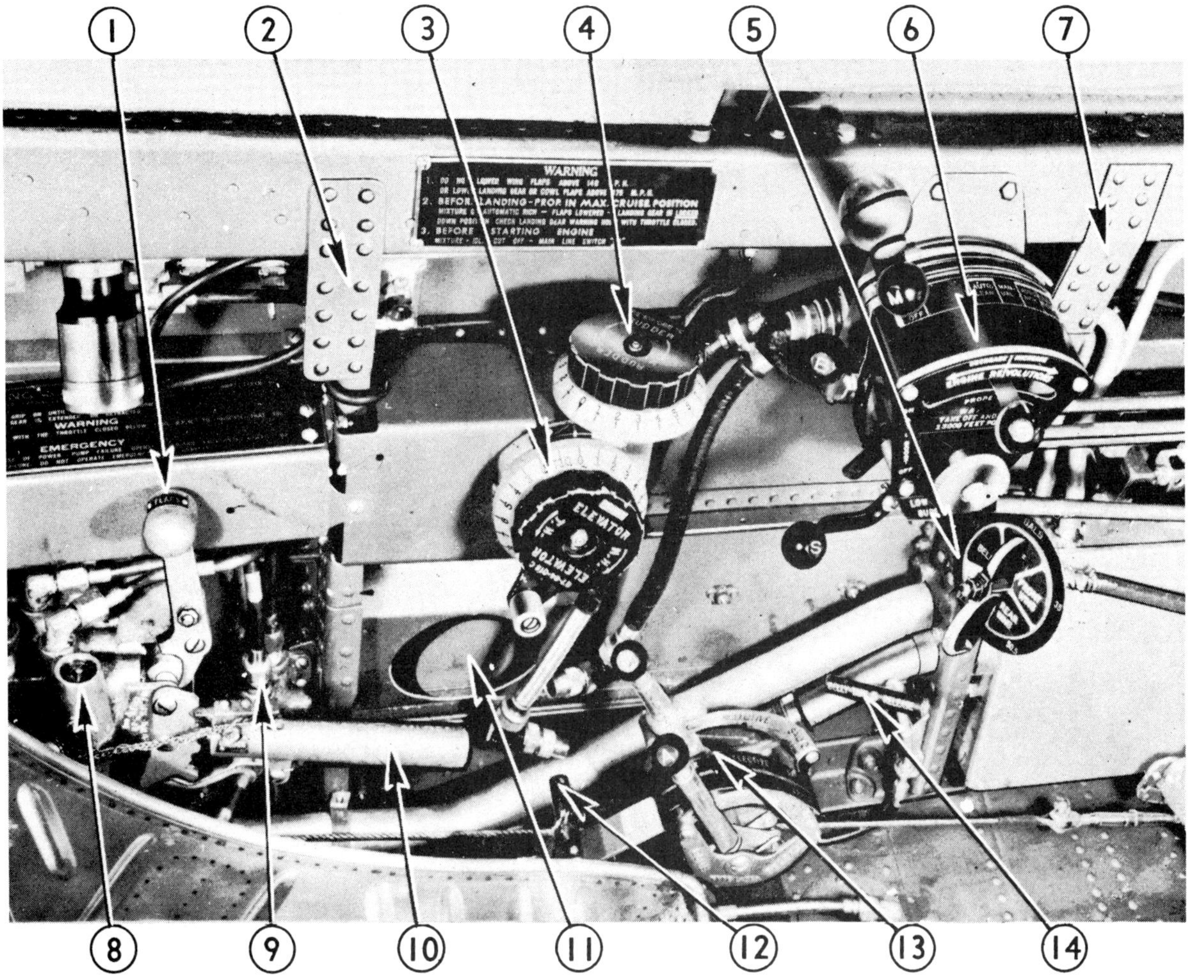

1. Flap control
2. Station No. 4
3. Elevator trim tab control
4. Rudder trim tab control
5. Fuel selector
6. Engine control quadrant
7. Station No. 3
8. Hydraulic control valve
9. Landing gear control lock
10. Landing gear control
11. Gun camera control bracket
12. Belly bomb arming handle
13. Wing bomb release quadrant
14. Belly tank release

Left side view of P-40 Cockpit.

CURTISS "WARHAWK" (P-40F)

Curtiss P-40F "WARHAWK" in flight. The "F" and "L" models were powered by Packard V-1650 Merlin engines.

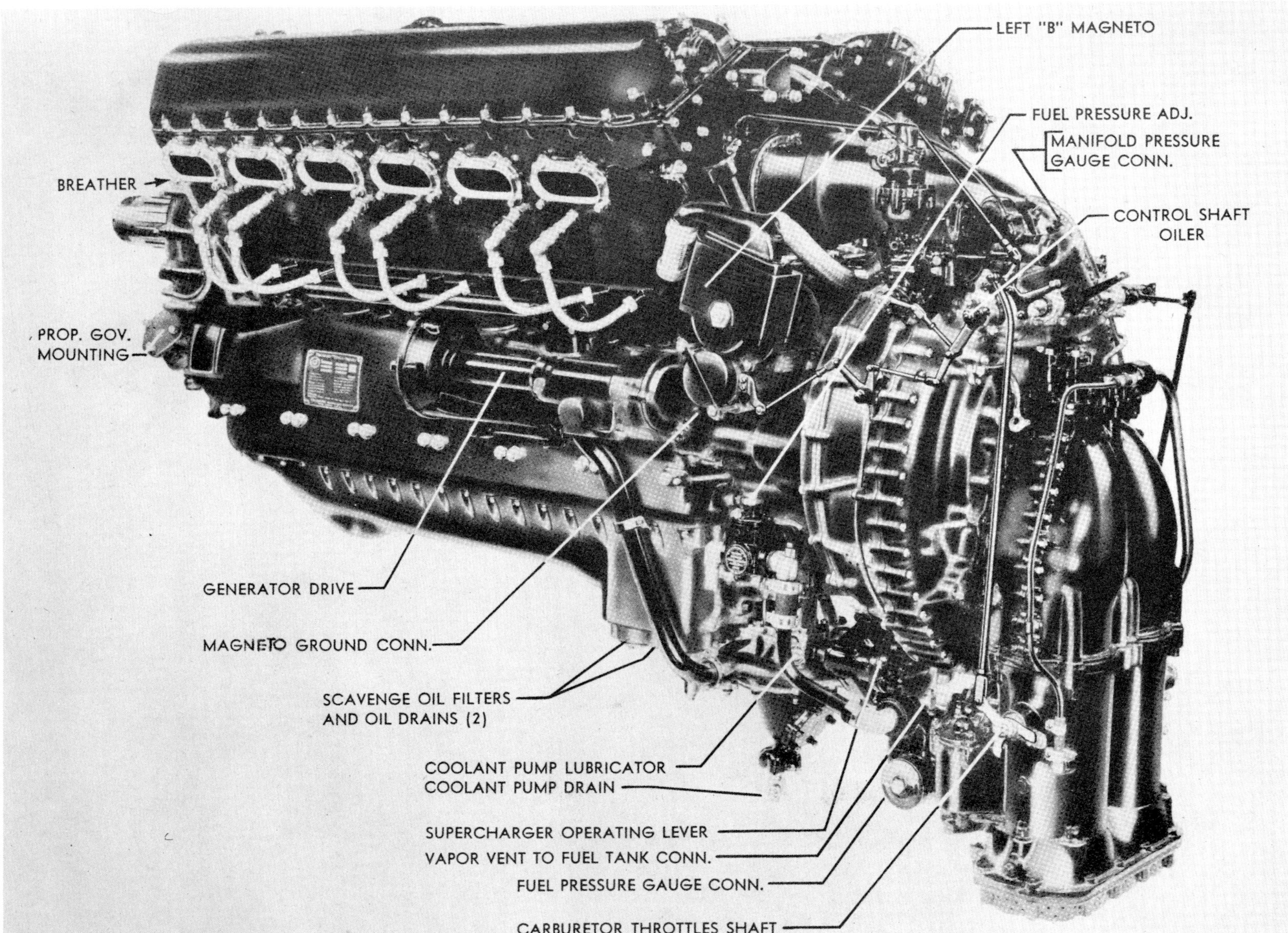

Three-quarter rear view of P-40 Allison V-1710 engine.

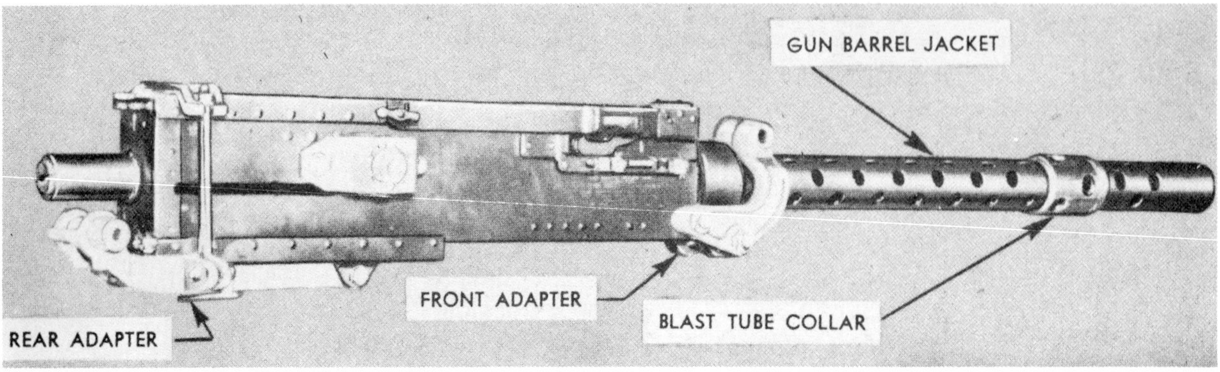

One of six Browning 50 cal. machine guns fitted to Curtiss P-40E.

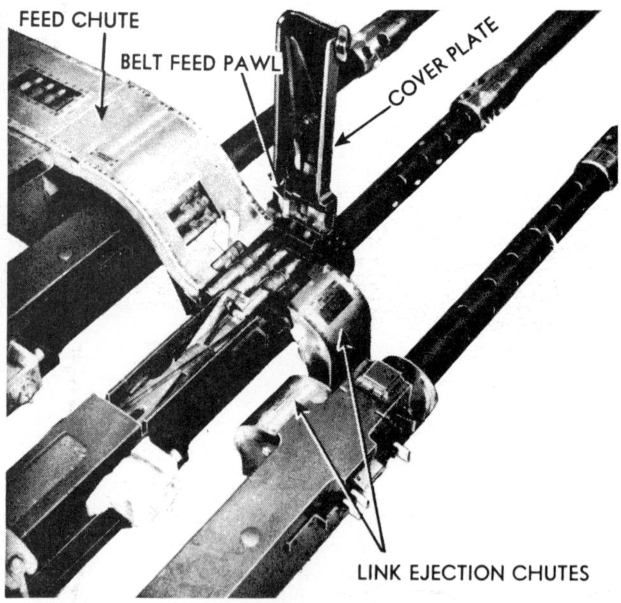

Three P-40 wing guns. Center gun is shown with belt — ready for firing.

Ammunition to P-40 wing guns were belt fed through chutes.

"Bombing Up" a Curtiss P-40. Sometimes referred to as the **"B-40"** in China-Burma operations.

This Civil P-40E lost its canopy in flight and was later fitted with a non-standard P-40N canopy. Note Fillet, baggage door and non-standard A.D.F. loop atop fuselage.

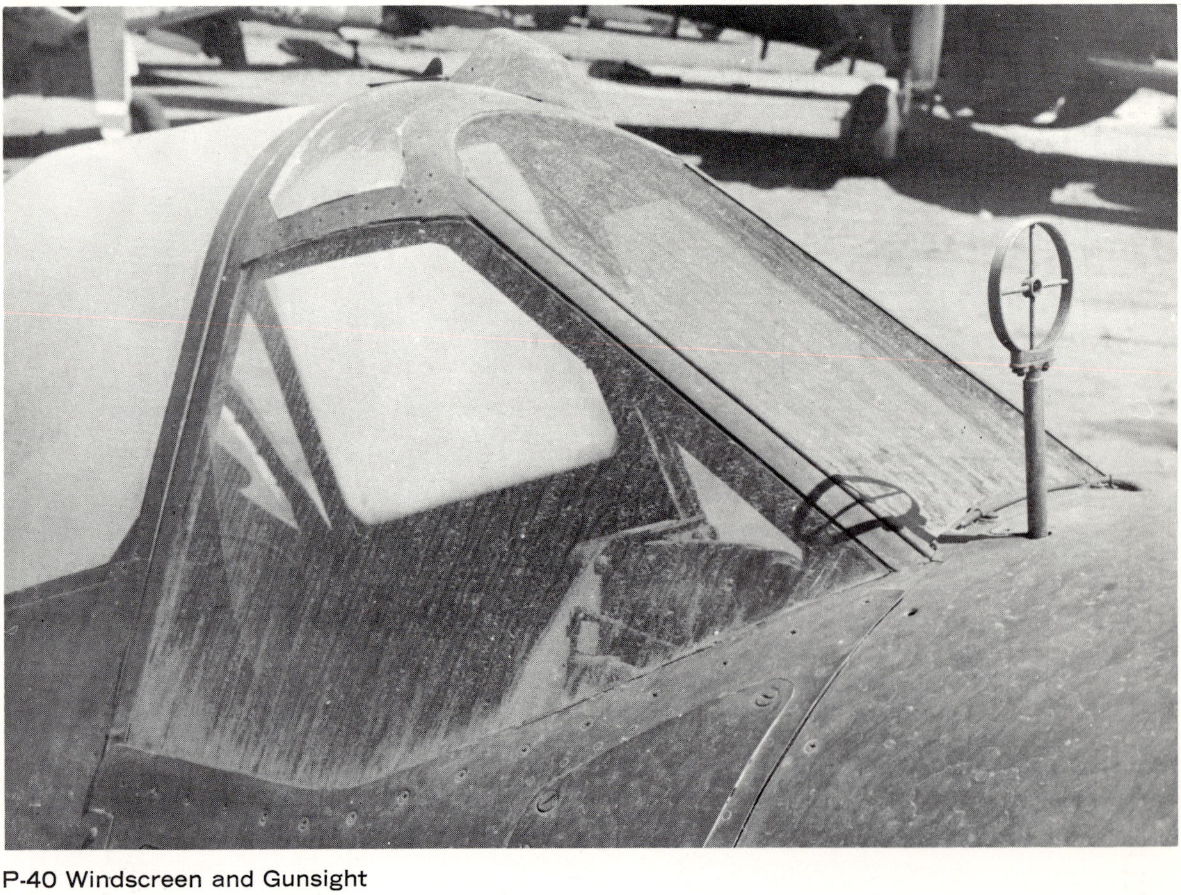

P-40 Windscreen and Gunsight

CURTISS P-40N-25-CU "WARHAWK"

The "N" model was one of the last production models of the P-40 series and was the light-weight fighter of the series. It was built in larger quantities than any other model and was the fastest example of all P-40's.

Engine:	Allison V-1710- 81, 1200 H.P.
	- 99
	-115
Speed:	375 M.P.H.
Armament:	6-50 cal. Browning M.G.
	3-500 lb. Bombs
Weight:	Gross, 9,500 lbs.
	Empty, 6,200 lbs.

P-40 Tail Assembly

Movable Tail Surfaces on all P-40's were of metal construction but fabric covered.

P-40 was one of the first of the Modern Fighters to incorporate a Retractable Tail Wheel.

P-40 Tail Wheel retracts into Rear Fuselage.

Curtiss P-40 in flight during acceptance test for U. S. Army.

Curtiss XP-40 — Three were built for Service Testing as no YP-40's were constructed.

Early production P-40B of the 20th Pursuit Group. Unfortunately none of the early Production P-40 B & C models can be located and must be presumed that all were scrapped.

CURTISS P-40 "WARHAWK" of famous Flying Tigers' 23rd Fighter Group — 14th Air Force

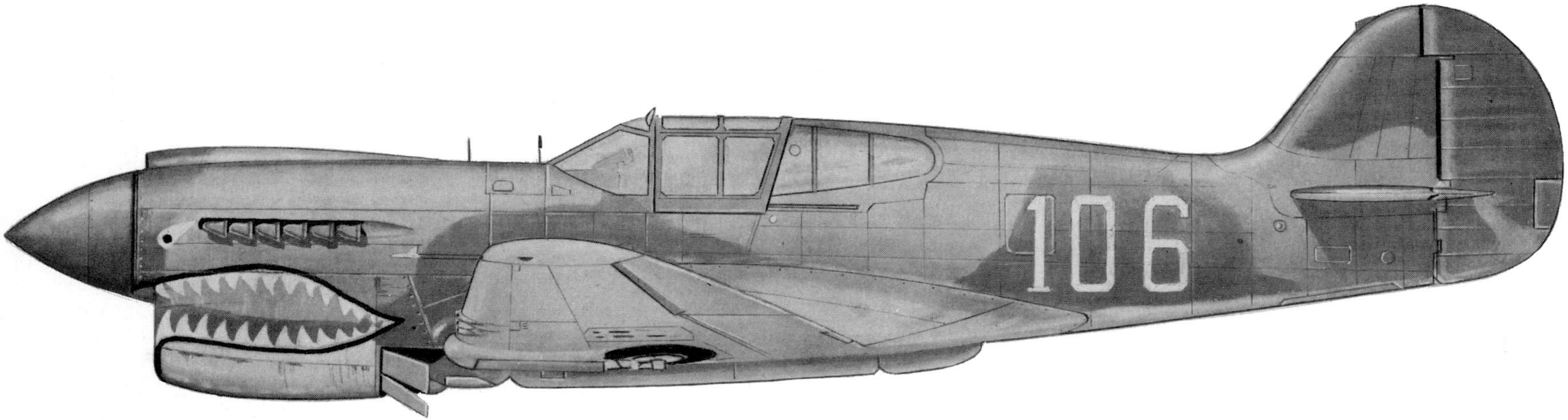

This model first appeared in August 1942 and was supplied to Gen. Chennault in large quantities. The P-40K model is easily distinguishable by the dorsal fin that was added to the rudder to correct a tendency to swing during take-off caused by a more powerful Allison Engine. An olive drab was applied over the brown of the upper surface. The individual Aircraft Number "106" was painted white and the Spinner red.

Engine:	Allison V-1710-73
Speed:	360 M.P.H. @ 15,000 feet
Armament:	6-50 cal. Browning M.G.
Weight:	Gross, 9,200 lbs.
	Empty, 6,350 lbs.

P-40F used "Merlin" engine. This one was used by 79th Fighter Group, 9th Air Force in North Africa.

P-40L also used a Merlin Engine. P-40F and L Models were only production models equipped with Merlin engines. All other P-40 Models had Allisons.

P-40M N1232N at San Fernando, California, in U. S. Civil Markings.

Curtiss XP-40 Q-3 42-45722 in Flight.

Curtiss XP-40 Q-2 had "Allison" V-1710-121 engine, square cut wing tips, and four-blade propeller and Bubble Canopy.

Curtiss XP-40Q-3 — This view clearly shows wing edge, oil and coolant radiators.

"AERO SHOP"

By Al Parker

Over the years there have been many model kits of the (
being the Aurora P-40E, Monogram P-40B in 1:48 scale, the F
fix P-40E and Monogram P-40N, all in 1:72 scale. Because c
dwell too long on any one kit but merely mention the good a
few suggested modifications.

Without a doubt, the finest P-40 kit produced is the Mo
Rivet detail, panel lines, general outline, all are accurate and
criticism is on the decal sheet. The color separations do not li
blue used in the Chinese insignia should have been lighter.

An interesting conversion of this kit would be the origin
These can be done by cutting off the radiator air scoop, filling
and body putty and adding the proper air scoops for the resp

Second-place honors must also go to Monogram for their
not as highly detailed as its larger brother, but still leads the fi
scale kits. For complete accuracy, the serious modeller may
stacks. This is done by sawing the exhaust panel out of a Rev
×15/16″ slot cut in the Monogram kit and cementing the stack

A simple conversion of this kit would be the P-40M. This
canopy to the N fuselage and filling in the headrest and fairin
wardly, this was the only difference between the two models.

For the truly ambitious modeller, the P-40N can be con
join the two halves of the fuselage. After the cement dries, cut
prop retainer and scoop ring for a future modification.) Now cu
hind the cockpit level with the root of the fin. Fill with scrap
the body putty is drying, cement the wing sections together, c
and round the corners slightly with sandpaper. At this point, yo
P-51 canopy to the cockpit and finish shaping the empennage.
of wood 1-3/16″ long, fit the exhaust panels from a Revell ki
the prop from the Lindberg P-51. The balance of the kit is st
cations need be done. Be sure to sand down all the rivet detail
form appearance. Overall finish is aluminum. If you take your ti
have an excellent P-40Q.

As for Revell, Airfix and Frog, these kits can all be classif
giving just a bit more detail. Revell, for example, is the only c
accurate landing gear, wheels and exhaust stacks. However,
a bit heavy, as is the canopy framing. Airfix, on the other hand
detail. They provide a finely detailed auxiliary fuel tank with bra
cast the wheel wells off center of the strut lines in the lower pa
eller who prefers no rivet detail at all, the Frog kit is it. Hov
national insignia, are cast into the surface of the plastic. Mar
my advice: buy all three and then pick the one you like best.

A rather simple conversion can be made from the standar
P-40E into a P-40F. Carefully file down and sand smooth the
cowling. P-40F's, which were Rolls-Royce Merlin powered, did

Last, but not least, is the Aurora P-40E. This is a good
about shape or fit; just practice with it.

Later on, we'll cover the Monogram P-40B conversion ir

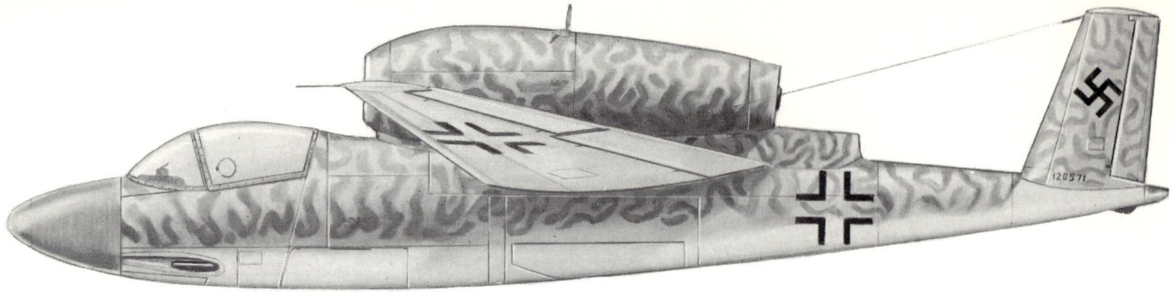

From the drawingboard to the initial flight tests in 70 days, this is the fantastic short period it took the Heinkel engineers to create the He 162 "Volksjager" (People's Fighter).

Photographs were taken directly from the Heinkel He 162, now on display at the Air Museum, showing closeups of the cockpit interiors, the BMW 003 A-1 turbojet, the 2-30 mm Mk 108 cannon, etc. Also included, to continue to give the book the outstanding quality of the AERO SERIES, will be side view and three-view color drawings.

Volume 5 will feature America's best known fighter of the 1930's—the colorful "Boeing P-12". One of these airplanes, in flying condition, can be seen at the Ontario Air Museum.

Many photos in this book have never been published before, and with the addition of five side views in color, this book promises to be a real gem and will be invaluable to model builders.